Cycles in Nature™

THE FOUR SEASONS

Suzanne Slade

The Rosen Publishing Group's

PowerKids Press™

New York

To Mrs. McNeir and Mrs. Novak, for helping the kids at Rockland School through every season

Published in 2007 by the Rosen Publishing Group, Inc.
29 East 21st Street, New York, NY 10010

First Edition

Editor: Joanne Randolph
Book Design: Greg Tucker
Photo Researcher: Amy Feinberg

Photo Credits: Cover, pp. 8, 14, 20, 26 © Jim Zipp/Photo Researchers, Inc.; pp. 4, 6, 8, 10, 14, 18 by Greg Tucker; p. 5 Courtesy of SOHO (ESA & NASA); p. 11 (top) © George McCarthy/Corbis; p. 11 (bottom) © Layne Kennedy/Corbis; p. 17 (top) © PunchStock; p. 17 (bottom) © Cath Mullen/Frank Lane Picture Agency/Corbis; p. 23 (top) © AP/Wide World Photos; p. 23 (bottom) © Ariel Skelley/Corbis; p. 29 (top) © Michael Siluk/The Image Works; p. 29 (bottom) © Mitch Wojnarowicz/Amsterdam Recorder/The Image Works.

Library of Congress Cataloging-in-Publication Data

Slade, Suzanne.
  The four seasons / Suzanne Slade.— 1st ed.
      p. cm. — (Cycles in nature)
  Includes index.
  ISBN 1-4042-3489-6 (lib. bdg.) — ISBN 1-4042-2198-0 (pbk.) — ISBN 1-4042-2388-6 (6 pack)
  1. Seasons—Juvenile literature. 2. Earth—Rotation—Juvenile literature. 3. Earth—Orbit—Juvenile literature. 4. Life cycles (Biology)—Juvenile literature. I. Title. II. Cycles in nature (PowerKids Press)
  QB637.4.S58 2007
  525'.5—dc22
                                    2005034872

Manufactured in the United States of America

Contents

Earth's Seasonal Cycle   4

Signs of Winter   6

Life in Winter   8

Signs of Spring   10

Life in Spring   12

Signs of Summer   14

Life in Summer   16

Signs of Fall   18

Life in Fall   20

Seasons Around the World   22

Glossary   23

Index   24

Web Sites   24

Earth's Seasonal Cycle

Most places on Earth have four seasons each year. These are winter, spring, summer, and fall. Every season has its own **temperature**, amount of sunlight, and weather patterns. Seasons last about three months. The seasons may be more or less noticeable depending on where you live.

All **planets** in our **solar system**, including Earth, move around the Sun. Earth orbits the Sun in about 365 days. As Earth travels around the Sun, it leans at an angle. That means different

AXIS
23.5°
EQUATOR

Earth leans, or tilts, at about a 23.5-degree angle, as shown in this picture. The orange line shows where the blue line would be if Earth had no tilt. Degrees tell us how far we have gone around a circle. A complete circle has 360 degrees.

parts of Earth are closer to the Sun at different times of the year. The part of Earth that is leaning toward the Sun receives more daylight hours and warmth than does the rest of Earth. Seasons are created by the amount of light and warmth that certain areas of Earth receive from the Sun during the year.

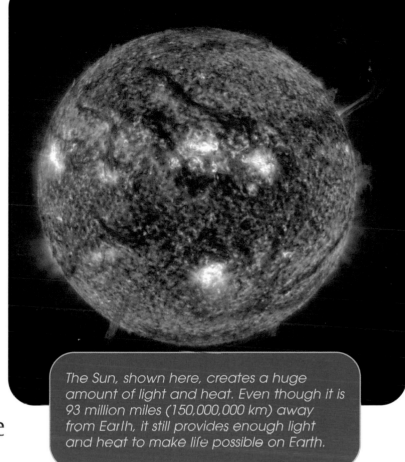

The Sun, shown here, creates a huge amount of light and heat. Even though it is 93 million miles (150,000,000 km) away from Earth, it still provides enough light and heat to make life possible on Earth.

Cycle Facts

Earth spins around its axis as it orbits the Sun. The axis is an imaginary line that goes through the middle of Earth. Every 24 hours Earth makes one complete turn, which is called a rotation. On the side of Earth that faces the Sun, it is daytime while on the side turned away from the Sun, it is nighttime.

Signs of Winter

The winter season is from mid-December to mid-March in the Northern **Hemisphere**. The Northern Hemisphere is north of the equator, which is an imaginary line that goes around the center of Earth. In winter temperatures are usually cold because the Northern Hemisphere leans away from the Sun. Earth's lean causes the Sun to rise later and set earlier, so days are shorter.

Although some people think winter arrives with the first snow, winter officially begins about

WINTER

This picture shows that Earth's Northern Hemisphere is leaning away from the Sun, which creates winter in that area. As the Earth moves around the Sun, the amount of sunlight that hits a certain location changes. This has an effect on how warm or cold it is in that region. This is why many areas of Earth have four seasons.

December 22 in the Northern Hemisphere. This day is called the winter **solstice**. It is the shortest day of the year. Snow, ice, and **freezing** temperatures are all signs of winter. It is winter in the Southern Hemisphere from mid-June to mid-September when it leans away from the Sun. In the Southern Hemisphere, the winter solstice is about June 21.

In winter many plants die. Other plants lose their leaves, as this maple tree has done. Winter often brings cold temperatures and ice or snow.

Cycle Facts

When the Northern Hemisphere is leaning away from the Sun in winter, the Sun stays low in the sky. The Sun's rays provide less warmth because they are traveling at an angle, rather than shining directly on Earth. This means they need to travel a longer distance through the air around Earth.

Life in Winter

The cold winter season brings changes for plants, animals, and people. For example, most plants stop growing and become dormant, or not active. Green, leafy **deciduous** trees drop their leaves and look like bare sticks. One kind of tree stays green during winter. This kind of tree is called an evergreen. Many evergreen trees have thin needles that do not fall off in cold weather.

Many animals and **insects** hibernate all winter. Some birds, such as swallows and robins,

Many animals hibernate during the coldest part of the winter. This dormouse is hibernating in its nest as it waits for warmer weather.

migrate to warmer areas. Animals that do not hibernate or migrate must search hard for food and water. People tend to stay inside more during the winter. Some people play sports that use winter's ice and snow, such as ice-skating and skiing. People also enjoy winter's holidays, like Christmas, Hanukkah, Kwanzaa, New Year's Day, and Valentine's Day.

Many people enjoy winter sports, such as sledding or ice fishing. Here a boy and his father ice fish in Wisconsin.

Cycle Facts

When animals hibernate their heartbeat slows down and their body temperature drops. Animals can hibernate for many months. Some ground squirrels stay in hibernation for eight whole months!

Signs of Spring

When spring begins, neither the Northern Hemisphere nor the Southern Hemisphere is pointed toward the Sun. As spring continues, the Northern Hemisphere leans a little more toward the Sun every day. This allows the Sun to appear a little higher in the sky each day. Days become longer and temperatures rise. The first day of spring arrives about March 21 in the Northern Hemisphere. This day is called the spring **equinox**. The word equinox comes from the Latin words for "equal" and "night." The number

SPRING

Temperatures start to rise in the Northern Hemisphere as the Earth continues its trip around the Sun. This is because the Northern Hemisphere points more toward the Sun each day.

of hours in the day and the night are equal on this day. The hard, frozen ground **thaws** in the spring and becomes soft and muddy. Signs of spring include melting snow, rain showers, longer days, and windy weather.

This maple tree has started to grow leaves once again. Spring brings warmer temperatures and often wet weather, which helps plants grow.

Cycle Facts

Toads lay small black eggs near the edge of ponds and lakes in spring. These eggs are warmed by the Sun, and tadpoles break free from them. Tadpoles look like tiny black fish. Between one and three months later, tadpoles lose their tails, grow legs, and become toads.

Life in Spring

The longer days and warmer weather of spring cause plants to start growing. Bare deciduous trees sprout fresh new leaves. Flowers, such as **daffodils** and tulips, break through the soft spring dirt. Bright green grass begins to grow on the ground. The rains of spring also help plants grow.

During the spring birds return from their winter migration. They build nests to lay their eggs. Animals, such as mice, bears, and raccoons, come out of hibernation. They are hungry after their long winter

This mother bird has a nest full of chicks to feed. The babies are usually ready to leave the nest after about two weeks.

This farmer is planting spring wheat. Spring wheat is planted in Montana, South Dakota, and Minnesota. It is called spring wheat because it is planted in the spring, instead of the winter, as wheat is in some areas.

and search for food. Butterflies, ants, and thousands of other bugs also reappear in spring. Spring is also the time when farmers begin to plant crops, like corn and wheat, in their fields. People also plant vegetable gardens and flower beds.

Cycle Facts

Plants that need to be planted every year are called annuals. Some annual plants produce flowers and vegetables during the summer, but these plants die when freezing temperatures arrive. Perennial plants do not die during winter. They lie dormant under the ground during the winter and come up each spring. Some examples of perennial plants are tulips and a green vegetable called asparagus.

Signs of Summer

As the Northern Hemisphere leans toward the Sun, it becomes summer in that area. Days become longer in the summer. Earth's northern half becomes hotter because it has more time to receive heat from the Sun. Temperatures also rise because of the Sun's position in the sky. The Sun moves high in the sky during the middle of a summer day. The Sun's rays travel a shorter distance through the air around Earth than they do in winter. This means more of the Sun's warmth heats Earth.

SUMMER

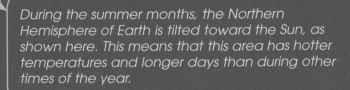

During the summer months, the Northern Hemisphere of Earth is tilted toward the Sun, as shown here. This means that this area has hotter temperatures and longer days than during other times of the year.

The hotter temperatures increase the amount of Earth's water that **evaporates** into the air.

The first day of summer is about June 21 in the Northern Hemisphere. This day is called the summer solstice. It is the longest day of the year. Longer days and hot temperatures are signs of summer.

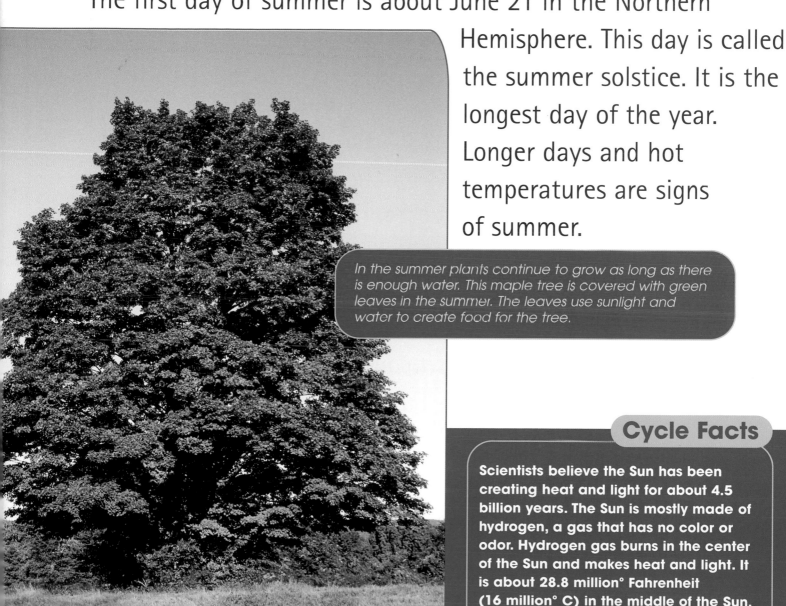

In the summer plants continue to grow as long as there is enough water. This maple tree is covered with green leaves in the summer. The leaves use sunlight and water to create food for the tree.

Cycle Facts

Scientists believe the Sun has been creating heat and light for about 4.5 billion years. The Sun is mostly made of hydrogen, a gas that has no color or odor. Hydrogen gas burns in the center of the Sun and makes heat and light. It is about 28.8 million° Fahrenheit (16 million° C) in the middle of the Sun.

Life in Summer

Nature is full of activity in the summer. Butterflies and bees are busy flying from flower to flower collecting food. At night the flashing lights from fireflies fill the air. Baby birds try their wings as they leave their nests for the first time. Vegetable gardens, fruit trees, and crops grow quickly in the hot sunlight. People enjoy eating fresh summer fruits and vegetables, such as watermelon and corn on the cob.

Summer is many people's favorite season. They enjoy the hot weather and long days. People

Many people enjoy doing outdoor activities, such as playing softball, during the warm days of summer.

Fireworks are a fun way to mark Independence Day. The United States fought against Britain to win its freedom and independence between 1775 and 1783.

spend a lot of time outdoors fishing, swimming, playing baseball, and having cookouts. Families often take vacations while children are out of school during summer. People in the United States **celebrate** their freedom and independence on the Fourth of July holiday with loud and colorful fireworks shows.

Cycle Facts

Although the Sun is about 93 million miles (150 million km) from Earth, you can still feel heat from its strong rays. The Sun's light and heat helps plants grow, but it can sometimes be harmful. You should protect your skin from burning by wearing hats and sunscreen lotion if you are out in the sun.

Signs of Fall

Earth completes its journey around the Sun from mid-September to mid-December. As fall begins neither the Northern Hemisphere nor the Southern Hemisphere of Earth is pointing toward the Sun. This is just like what happens in spring. As fall continues, though, the Northern Hemisphere starts to lean away from the Sun. The Sun rises later and sets earlier each day. The days become shorter, the Sun does not climb as high in the sky, and temperatures drop.

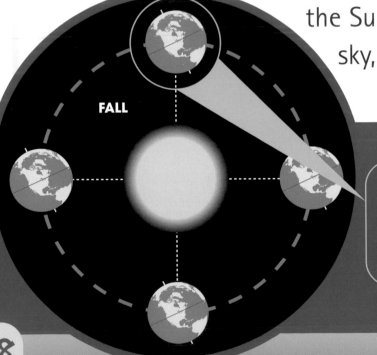

FALL

Here you can see where Earth and the Sun are positioned in the fall season. In early fall the days are still warm and long. As the season continues, though, the days get shorter and the air becomes cooler. Soon it will be winter again, and the cycle will continue.

The fall season begins about September 22 in the Northern Hemisphere. This day is called the fall equinox. Day and night are both 12 hours long on the fall equinox. Cool days, colorful falling leaves, and children going back to school are all signs that fall has arrived.

This maple tree's leaves have changed color. This is a sign of the fall season. Leaves stop producing food and lose their green color.

Cycle Facts

When the Northern Hemisphere is having its fall season, it is spring in the Southern Hemisphere. The spring equinox in the Southern Hemisphere is around September 22, the same day as the fall equinox in the Northern Hemisphere. On this day neither hemisphere is leaning toward the Sun, so they both get light from the Sun for the same amount of time.

Life in Fall

Fall is called a season of **harvest** because farmers spend long hours bringing in crops from their fields. Corn, apples, and pumpkins become ripe in fall. Animals gather and store food for the cold winter ahead. Animals also eat a lot of food to put extra fat on their bodies. They grow a thick coat of warm fur, too. Some birds begin to migrate to warmer places.

Green leaves on deciduous trees turn bright colors in fall. During summer these leaves are green because they have **chlorophyll**. Trees use

These boys are raking leaves that have fallen from the trees.

chlorophyll and sunlight to make food using **photosynthesis**. In fall there is less sunlight, so leaves stop making food and chlorophyll. When the chlorophyll in leaves fades, other colors that were hidden by the green chlorophyll can be seen.

Apple harvesting is another sign of fall. Here a farmer picks apples on his orchard in New York. An orchard is the name for a group of fruit or nut trees.

Cycle Facts

The leaves of different deciduous trees turn certain colors. For example, the leaves on a tree called a maple turn red, yellow, and orange. Oak tree leaves become brown and red in fall.

Seasons Around the World

Some places on Earth do not have four seasons. For example, the **tropics** have two seasons, dry and rainy. The tropics are near the equator. This area does not lean toward or away from the Sun but is close to the Sun all year long. The North Pole and the South Pole, the ends of Earth farthest from the equator, also have two seasons. The weather is always cold during these seasons, called polar summer and polar winter. In polar summer the Sun stays in the sky for six months. The Sun disappears for a long six-month night during polar winter.

The seasons you have depend on where you live. Changing weather is the most noticeable change in each new season. People, animals, and plants also change during each season of Earth's seasonal cycle.

Cycle Facts

Planets such as Uranus and Mars have seasons, too. These planets lean at an angle as does Earth. The more a planet leans toward the Sun, the more noticeable the winter and summer seasons will be on that planet. Mercury does not have seasons because it does not lean toward the Sun at all.

Glossary

celebrate (SEH-luh-brayt) To observe an important occasion with special activities.

chlorophyll (KLOR-uh-fil) Green matter inside plants that allows them to use energy from sunlight to make their own food.

daffodils (DA-fuh-dilz) Bright yellow flowers that come up in the spring.

deciduous (deh-SIH-joo-us) Having leaves that fall off every year.

equinox (EH-kwih-noks) One of the two days of the year when the hours of day and night are equal.

evaporates (ih-VA-puh-rayts) Changes from a liquid to a gas.

freezing (FREEZ-ing) Changing from a liquid to a solid.

harvest (HAR-vist) A season's gathered crop.

hemisphere (HEH-muh-sfeer) One half of Earth or another sphere. A sphere is something shaped like a ball.

insects (IN-sekts) Small creatures that often have six legs and wings.

migrate (MY-grayt) To move from one place to another.

photosynthesis (foh-toh-SIN-thuh-sus) The way in which green plants make their own food from sunlight, water, and a gas called carbon dioxide.

planets (PLA-nets) Large objects, such as Earth, that move around the Sun.

solar system (SOH-ler SIS-tem) A group of planets that circles a star.

solstice (SOL-stes) The longest or shortest day of the year.

temperature (TEM-pur-cher) How hot or cold something is.

thaws (THAHZ) Turns from solid to liquid, such as when ice and snow melt.

tropics (TRAH-piks) The warm parts of Earth that are near the equator.

Index

B

birds, 8, 12, 16, 20

C

chlorophyll, 20

D

deciduous trees, 8, 12, 20

E

Earth, 4–6, 14–15, 18, 22

F

fall equinox, 19

flower(s), 12, 16

H

harvest, 20

holiday(s), 9, 17

N

Northern Hemisphere, 6–7, 10, 14–15, 18–19

P

photosynthesis, 21

S

solar system, 4

Southern Hemisphere, 7, 10, 18

spring equinox, 10

summer solstice, 15

Sun, 4–7, 10, 14, 18, 22

T

temperature(s), 4, 7, 10, 14, 18

tropics, 22

W

winter solstice, 7

Web Sites

Due to the changing nature of Internet links, PowerKids Press has developed an online list of Web sites related to the subject of this book. This site is updated regularly. Please use this link to access the list:

www.powerkidslinks.com/cin/seasonal/